Preparing Kids for Math Facts Fluency

Help Your Child Develop Basic Math Skills with Simple Activities at Home

David Itanola

No part of this publication may be reproduced, or transmitted in any form or by any means including photocopying, recording or other electronics or mechanical method without the prior permission of the publisher except for brief quotations embodied in critical reviews.

Copyright © 2017 David Itanola
ISBN-13:978-1977649980
ISBN-10:197764998X

Table of Contents

An Overview of this Book   iv

Introduction   v

1. Counting and Number Recognition   1

2. Developing Math Facts Fluency   12

3. Strategies for Addition and Subtraction Facts   15

4. Teaching Doubles and Near Double Facts   25

5. Number Sense Worksheets   43

    About the author   60

An Overview of this Book

This book is divided into 5 chapters arranged systematically to reinforce critical skills necessary for kids to succeed in kindergarten and first grade math.

We begin with Counting and Number Recognition activities. Then we proceed to strategies for developing Math facts fluency and conclude with Number Sense Worksheets that enhances number manipulation fluency.

Introduction

Promoting the development of basic number concepts before school entry sets the foundation for more advanced learning of math. It equips kids with the necessary confidence crucial to maintaining a positive attitude toward learning higher order Math.

With this book, learning math becomes easy for the entire family. You will be able to help your child to discover just how exciting numbers can be! Your child will be able to recognize numbers immediately and soon start counting with ease.

1. Counting and Number Recognition

Number literacy is a fundamental skill that a child must master for success in later math and science education. A child must be able to identify and recognize numbers before learning basic math skills such as addition and subtraction. Using age-appropriate techniques will help a child to master these essential math milestones with ease.

Number Sense

Number sense gives a child the ability to count accurately and see the relationship between numbers. It also includes basic math operation; addition, subtraction, multiplication, and division at a later stage. Here are some activities that give kids opportunity to gain the ability in a fun and playful manner.

Using math concept to illustrate the situation in the home

You can encourage kids to sing songs and recites rhymes that involve counting. E.g., one, two buckle my shoe and other related ones. You can also help him to memorize some important numbers such as your phone numbers and the number of your house.

Furthermore, you can adopt speaking in a language that emphasizes the use of time. For example, "it is 4 pm now, when it is 7 we will have our dinner and go to bed at 8 pm". This is a good way of letting a child know that dinner comes before bedtime. It will aid his understanding of the sequence of events and pattern in daily activities.

Cooking activities

Cooking activities at home can equally be used to teach number sense. For instance, when you are cutting, tell your child 'I am cutting this loaf into two.' Later you can allow the child to perform the exercise on his own; and let him tell you what he is doing. This practice will create the awareness of the existence of other numbers, and aid the understanding of the concept of the fraction at a later stage of learning.

With constant practice, the child will be able to name all the digits from 0-9 in no distant time.

Number Recognition

The next step in number recognition activity after rote memorization is helping kids to know what the first ten digits (0-9) look like. This ability requires strong visual discrimination because of the seemingly identical nature of some of these digits. At the early stage, it is easier for a child to confuse 6 for 9 or 1 for 7 and vice versa. Once a child can recognize these digits, he can be guided to develop an understanding of the actual value of each number with the appropriate quantity of objects. Note that using age-appropriate techniques will promote skills development rather than conceptual learning.

SYMBOL	NAME	OBJECTS
0	ZERO	-
1	ONE	🍎
2	TWO	🍎 🍎
3	THREE	🍎 🍎 🍎
4	FOUR	🍎 🍎 🍎 🍎
5	FIVE	🍎 🍎 🍎 🍎 🍎

Number Recognition Activities

The following simple activities are planned to aid quick recognition of numbers. The parents are to assist the child in the exercise

1. Cardboard Cut-outs

Cut cardboard into six smaller cards to represent the six different faces of a die. Indicate on each card the digits one through six and the corresponding number of dots.

Procedure

Let the child cast the die severally on his own. After each throw let him pick the corresponding card that represents the outcome then count the dots and say the number. For example, if "3" is the throw. Let him pick the card with three dots and the digit '3'. Ask him to count the dots and say "three." Let him do it continuously to get all the digits 1 through 6. With constant practice, the child will be able to recognize the appropriate symbol that represents each digit from one to six within a short time.

2. Number board

A number board is a rectangular wooden or plastic cutout board with numbers stacked on its surface. The numbers usually range from 0-9 and designed in such a way that each number contains the corresponding number of pictures to enhance counting and matching of objects. For example, when you remove the number 5 "you may see five pictures of objects say bags, behind the number."

Let your child count the objects and say five bags. Guide him to repeat the exercise for all the numbers until he can recognize the digits 0-9.

3. During a road journey, you can direct the child to point and name visible numbers on street signs and other objects on the road that contain numbers.

4. Ask the child to find numbers around the house in some items such as calendars, charts, telephones, calculator, etc. Let him interact with numbers and ask him to name the digit that represents his age. Similarly, make available a few sets of magnetic numbers and allow the child to play with them freely. You can ask him to match up pairs of the same number and put them in order.

5. Read counting books to your child; counting books contain numerals and the corresponding number of pictures on each page. In selecting a counting book, make sure the pictures are colorful, engaging and easy to count with

numerals printed boldly. As you read the counting book, encourage the child to say the numeral loudly on each page and count the object on the page as you touch each picture.

2. Developing Math Facts Fluency

Nowadays, attention should be shifted from memorization of facts and procedure to increased understanding of math skills and concepts.

Fluency in math facts is the ability to recall and apply basic math operations [addition, subtraction, multiplication, and division] effortlessly and accurately. The skill helps the child to feel comfortable solving math and conveniently give answers to some simple math problems.

Why is it important for kids

Fluency in Math Leads to Less Confusion and Anxiety

Fluency enables kids to consecrate on problem-solving and acquire more relevant skills. It reduces pressure on the working memory while solving problems. When a child masters math facts, he will be better equipped to solve problems faster. He will be able to devote more time thinking on the challenge at hand rather than struggling to recall necessary steps towards solving a problem. If a child spends a lot of time thinking about the basic facts, he is more likely to be confused with the processes and get lost in problem-solving process.

Math Facts Fluency Leads to Higher Order Math

Helps kids to develop a solid foundation for higher order math concept and reduces confusion and anxiety while solving a math problem. It enables a child to stay focused during problem-solving instead of spending most of the time figuring out the simple calculation which can lead to confusion during working. Furthermore it strengthens kid's abilities in solving math problems and prepares them for difficult challenges in the future math activities.

3. Strategies for Addition and Subtraction Facts

Addition facts involve the combination of single digits addends to give a sum not greater than 18. Examples are

2+5=7, 7+3=10 and 9+9=18

13+10= 23 is not a fact in this category because the addend is not a single digit number and the result is greater than 18.

Subtraction facts are the inverses of addition facts since addition and subtraction are related. Subtraction is the inverse operation of addition. 12-7=5 can be written as 12=7+5. Similarly, 16= 7+9 is same as 16-7=9.

Therefore 12-7=5, 16-7 =9 are examples of subtraction facts.

Facts Family Table

1+1	2+1	3+1	4+1	5+1	6+1	7+1	8+1	9+1
1+2	2+2	3+2	4+2	5+2	6+2	7+2	8+2	9+2
1+3	2+3	3+3	4+3	5+3	6+3	7+3	8+3	9+3
1+4	2+4	3+4	4+4	5+4	6+4	7+4	8+4	9+4
1+5	2+5	3+5	4+5	5+5	6+5	7+5	8+5	9+5
1+6	2+6	3+6	4+6	5+6	6+6	7+6	8+6	9+6
1+7	2+7	3+7	4+7	5+7	6+7	7+7	8+7	9+7
1+8	2+8	3+8	4+8	5+8	6+8	7+8	8+8	9+8
1+9	2+9	3+9	4+9	5+9	6+9	7+9	8+9	9+9

How to Teach the Addition Facts

Don't overwhelm your child with all of the addition facts at once; first break the facts into smaller groups and focus on this at the initial stage. You can first introduce teach+1 and +2 facts before moving on to other facts.

Counting up Strategy

This strategy is suitable to teach kids addition of one digit number. It will be easier when a child has mastered counting and number recognition up to ten.

Give your child three objects say apples, let him count and say 'I have three apples.

Ask him to display the three apples on the table.

Give him two more apples and ask him to merge the apples and count together.

Ask him 'how many apples do you have.

Three apples and two apples give five apples

3+2=5

5+2=7

This simple idea can aid kids to carry out single digits addend easily

For a quicker result, you can help kids to follow these steps in carrying out simple addition.

For example, **5+2**

Step 1: Ask the child to hold the bigger number which is 5 in his working memory.

Step 2: Ask him to start counting up 2 from 5, i.e. (5), 6 7 the two count ends at 7 which means 5+2 = 7

Another example

4 + 3

Step 1: Ask a kid to hold 4 in his working memory

Step 2: Ask him to start counting up 3 from 4, i.e. (4), 5, 6, 7, the counting terminates at 7 which means

4+3 = 7.

You can give more examples to help strengthen proper understanding of this concept and remember that this method works better when the addends are a single digit.

You can practice these with your kid

1. 5+2   6. 4 + 1

2. 7+1   7. 5+3

3. 6+3   8. 8+1

4. 4+3   9. 7+3

5. 7+2   10. 5+4

Subtraction Strategy

Important fact for subtraction of numbers

Let us consider the equation

9-7=2.

The number 9 is referred to as the minuend, while 7 is the subtrahend and 2 is the difference.

This fact is useful when subtracting numbers.

The difference of two numbers can be calculated by counting back or counting up

Counting Back

Counting back method is suitable when the subtrahend is small say 1, 2 or 3.

For example, let us look at 7-2=5 you can start by asking the child to start from 7 and count back 2 steps. This ends at 5.

This can be illustrated on a number line for proper understanding.

Counting up method is practiced by counting up from the least number to find the difference between the two given numbers. However, this method is only suitable when the minuend and subtrahend are closed together.

For example, 7-2 gives 5. This means how many counting up from 5 makes 7, which is 2.

Another good method is to rewrite subtraction in the form of addition.

9-6 can be written as ☐ **+ 6 = 9 or**

Or   6 + ☐ = 9   i.e. what adds up to 6 to make 9.

You can practice the following with the kids

1. 6-2= ☐

2. 5-4= ☐

3. 7-6= ☐

4. 9-8= ☐

5. 7-1= ☐

4. Teaching Doubles and Near Double Facts.

1+1=2

2+2=4

3+3=6

4+4=8

5+5=10

6+6=12

7+7=14

8+8=16

9+9=18

Double facts are obtained when two numbers with the same values are added together. These facts are easy to remember and can be used to promote number fluency for kids when dealing with small numbers usually less than twelve.

Double facts are relatively easy to learn because all the sums are even and form a counting by two patterns.

Near Double Facts

The idea of double facts can be extended to near double addends. Since the results of double facts are usually easy to remember, for example, 3+3 gives 6.

This fact can be used to teach or develop fluency for numbers close to 3.

For example, to solve **2+3**

With the result of the double facts of 2 in mind

First see 2+3 as (2+2) +1. I.e. double facts of 2 plus 1. Which is just one more than the double facts of 2.

Hence 2+3 = (2+2) +1 =5.

This method can be similarly employed to treat all numbers that are close to any double facts.

For example, 4+5 can be treated as (4+4) +1which is just adding 1 to the double fact of four. Other examples are 6+7, 7+9

In general, the strategy is to bring out the double facts in the given exercise where possible and add the remainder to give the answer.

Using 10 facts

Addition of numbers involving 10 is usually simple. When 10 is added to any number, it only affects the digit on the tense side. For example, 10+1 = 11,

10+2=12.

Other examples

10+4=14

10+6=16

10+8=18

This is valuable information in dealing with numbers that are very close to ten

For example to solve 18+9 and 18+11

First treat 18+9 as 18+10 =28. Now since 9 is 1 less than 10 then take away one from the result that is 28 - 1 = 27.

With this technique, the result of 18+9 can be given effortlessly.

Similarly, 18+11 first treat it as 18+10 =28 and add 1 to the result since 11 is just 1 greater than 10.

Constant practice of this strategy will help the child to master addition and subtraction of numbers.

This game is designed for preschoolers and kindergarten to strengthen the concept of number facts.

Activity; "Fill the basket."

Aim: Promoting Addition Facts of Ten.

Materials:

Balls (10)

Baskets (2)

Participants: 2 children

Step 1:

In an open space, mark two points X and Y about 10m - 20m apart. At point X place two different empty baskets one for each participant and let each child identify his basket properly before you start the activity. At point Y leave the 10 balls scattered on the floor with participants standing at them. Proceed by counting the balls together with the participants and let them.

Step 2:

Explain the process; let them know that they are to fill their baskets in X by running to Y to pick a ball at a time.

Step 3:

At the sound of the word start the participants are to pick a ball at point Y where they are standing and run to the other end to fill their baskets. They are to do this until the all the balls in Y are picked.

Step 4:

When there are no more balls, ask the children to empty their baskets and count the number they can collect from the 10 balls.

Step 5:

The child with the highest balls is the winner of the game. Use this exercise to explain the additional fact of 10 to the children by adding the results of their individual efforts.

Let them see clearly that no matter the individual effort in the game, their efforts will always add up to give 10.

You can influence the outcome of the game to bring out all the addition facts of 10 such as

10+0,

9+1,

8+2,

7+3,

6+4

5+5.

You can twist this game to teach other addition facts, just increase the number of balls to the fact you want to teach.

Teaching subtraction

This game can be extended to teach subtraction of numbers.

Materials:

Balls (10),

Basket

Stopwatch

Aim: Teaching subtraction facts of 10.

Description: At two different locations X and Y about10- 20m apart. At point X place the basket and let the 10 balls be on the floor at point Y.

Duration: 2-5 minutes

Step1:

Let the child stay with the balls at Y. Then at the sound of the word 'go' he is to empty all the 10 balls at Y into the basket at X by picking one ball at a time. He is to do this for the duration of the game.

Step2:

Let the child understand the game properly; the parent is to supervise and set the alarm for the duration of the game. The child is to stop at the sound of the alarm.

Step3:

At the end of the game, count the balls at both ends. If the child can pick only 2 balls for the duration of the game, ask him to tell you the number of balls at X in the basket and the number of balls at Y on the floor. Use this to teach problems such as **10-3** by asking the child 'if 3 is taken away from 10 what is the answer? The child will be able give the answer effortlessly.

You can continue to twist this game for the child to bring out all the facts.

10-0=10

10-9=1

10-8=2

10-3=7

10-8=2

Continuous practice of this activity will help the child to memorize addition and subtraction facts easily.

Facts Family

A group of numbers connected together by a certain order forms a number family. Math fact family equally begins with two numbers coming together with a math operation to produce the third number. This math operation could be addition and subtraction or multiplication and division. Let us say, for example, 2 and 3 come together also, the union will produce;

2+3=5 and 3+2=5 (commutative property of addition)

Similarly, we have 5-2=3 and 5-3=2.

Thus, the three numbers (2, 3, and 5) constitute a family.

Other examples are (4, 5, and 9)

4+5=9, 5+4=9 and 9-4= 5 and 9-5=4

Set of facts family

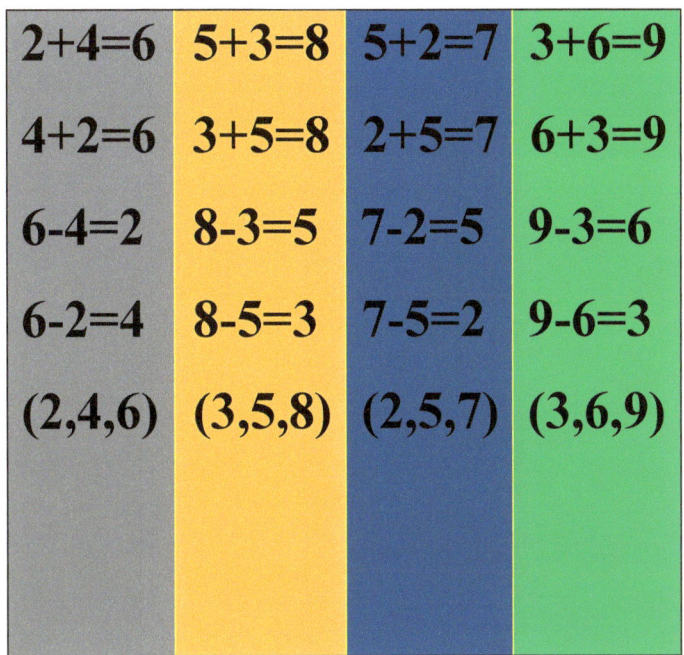

Facts family for Multiplication and Division

When multiplication and division are involved, we can similarly obtain a family

Let us consider 2 and 3, when the two numbers are multiplied we get the third number which is 6.

2x3=6 and 3x2=6 (commutative property of multiplication)

Conversely 6\2=3 and 6/3=2.

Thus (6, 3, and 2) is a fact family.

We can see clearly from the examples given that when two numbers are added or multiplied the sum or product makes a family with the two numbers.

Once a fact is known you can use this to work out the other facts to simplify your work. For instance, to get the missing number in these equations

2 + ☐ = 5 and

5 - ☐ = 3. The third number is 2

The child will be able to give the answer effortlessly if he is familiar with the fact family 2, 3, 5

Exercises

Find the missing numbers

1. 6 + ☐ = 8

2. 5 + ☐ = 9

3. 4 + ☐ = 6

4. 7 + ☐ = 8

5. 2 + ☐ = 7

6. 6 + ☐ = 7

7. $5 + \square = 7$

8. $4 + \square = 8$

9. $7 + \square = 9$

10. $5 + \square = 6$

5. Number Sense Worksheets

1 Complete the table below

1	2	3	4	5	6	7	8	9
		3			6			9
1			4				8	
	2			5		7		9
			4				8	

2. Shade the correct number of squares

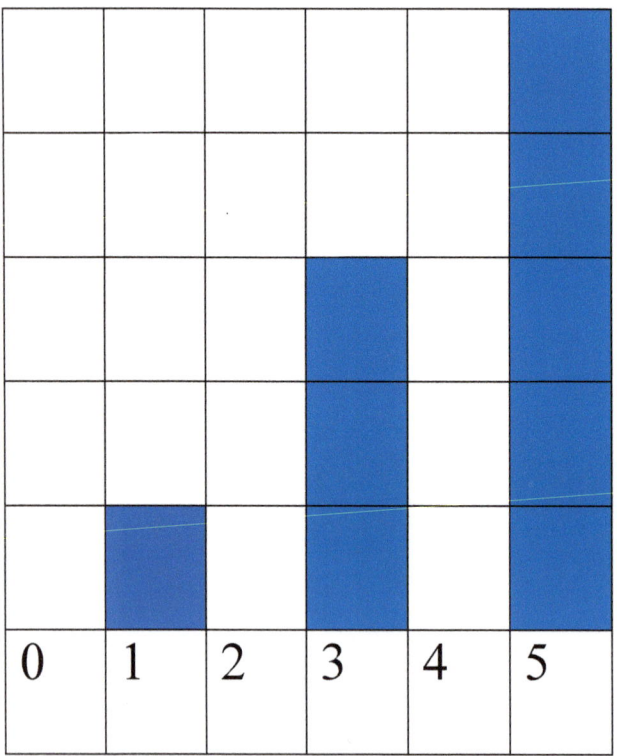

3. Complete by drawing correct number of circles

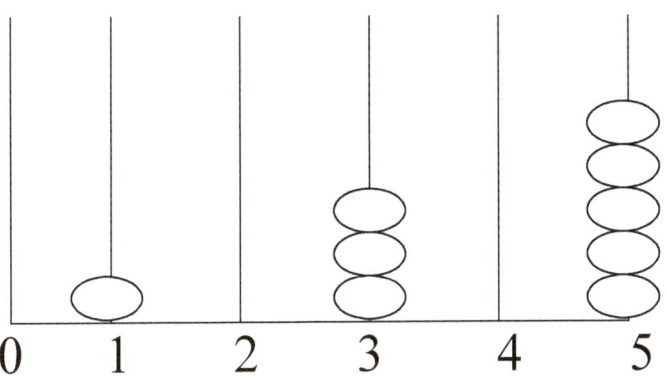

4. Shade the bigger Number
Example

i.

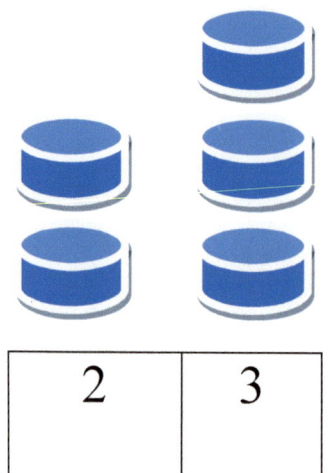

2	3

ii.

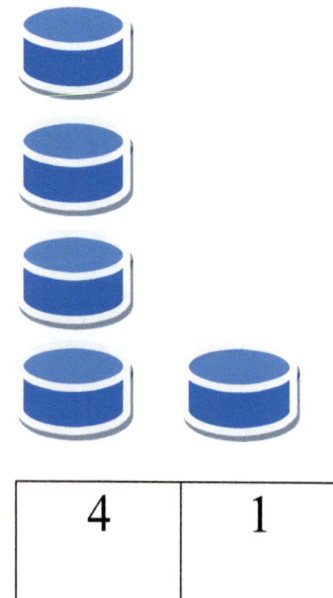

4	1

iii

4	0

iv.

4	3

v.

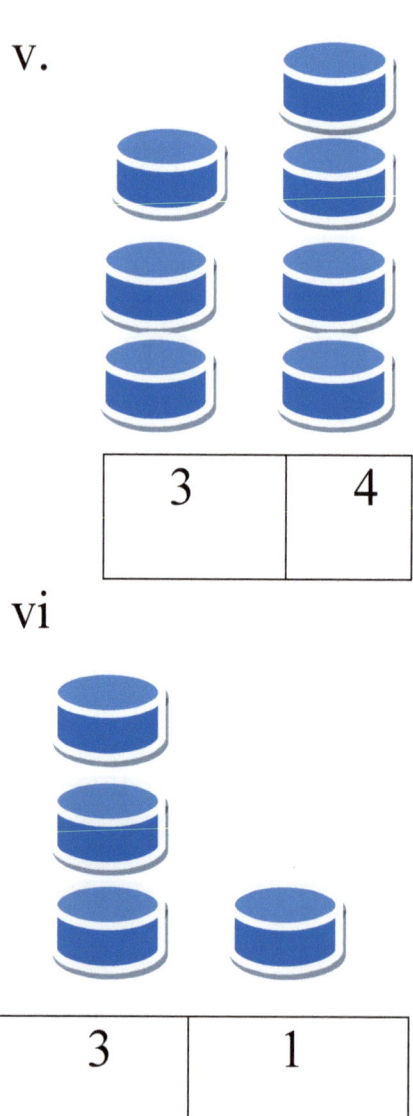

| 3 | 4 |

vi

| 3 | 1 |

5. Shade the smaller number

Example

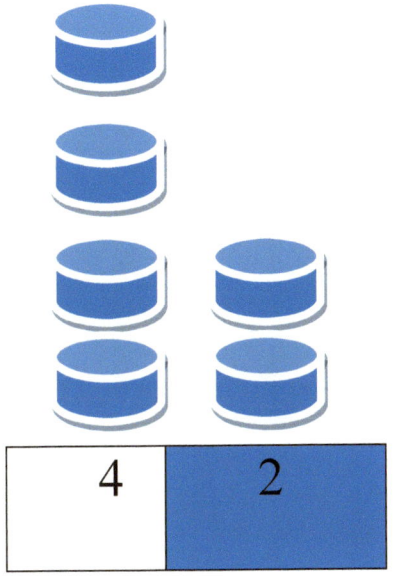

i.

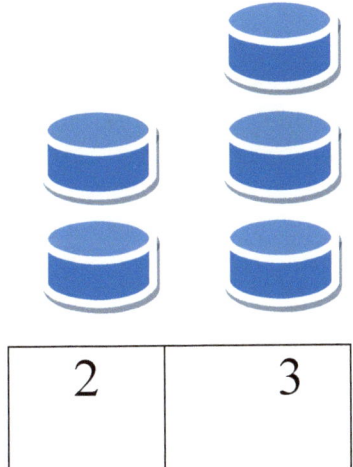

ii.

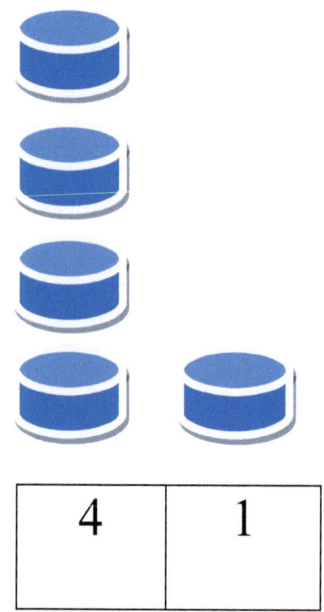

iii

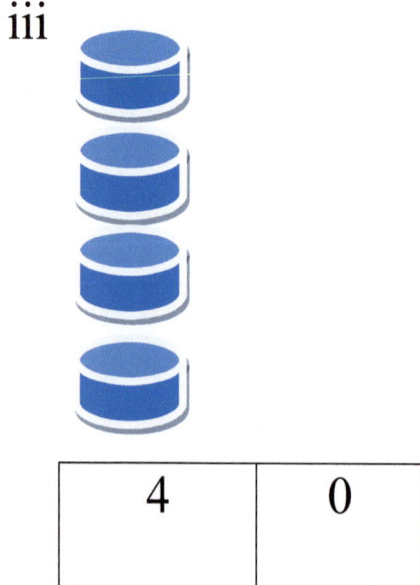

iv.

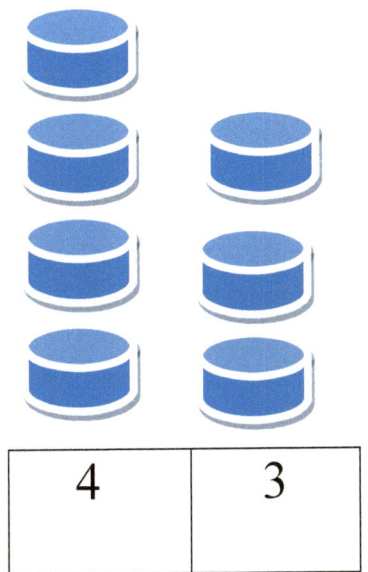

| 4 | 3 |

v.

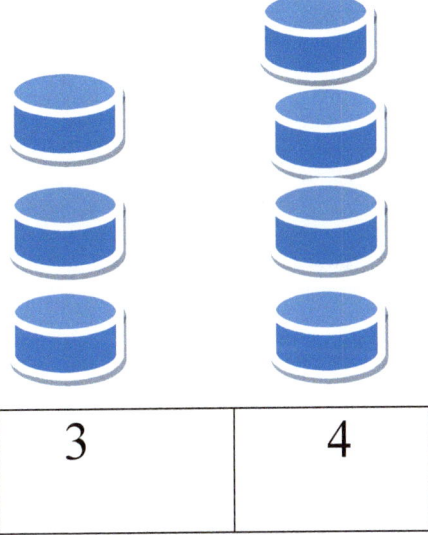

| 3 | 4 |

vi

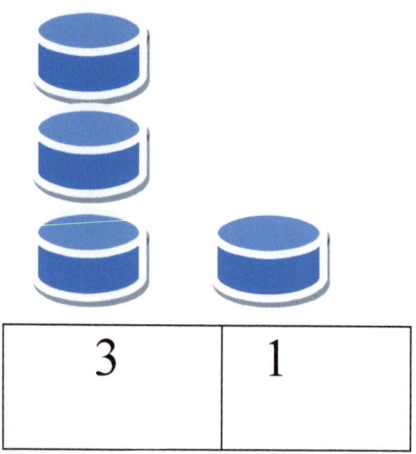

6. Count and Write the Number of Stars

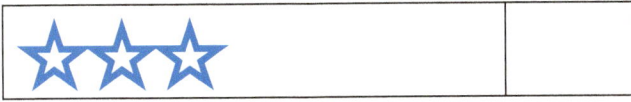

7. Tick set of 5 stars

Example

 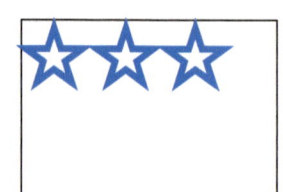

 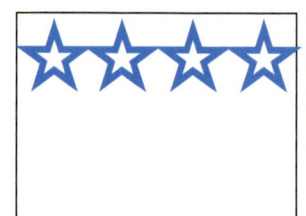

✓

i

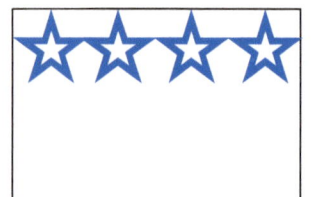 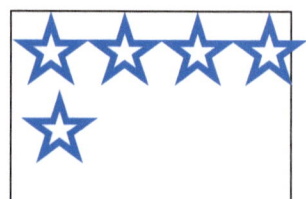

ii

iii.

 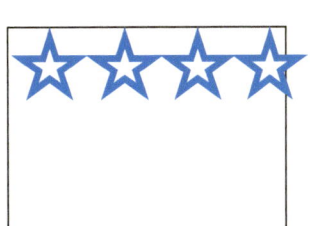

 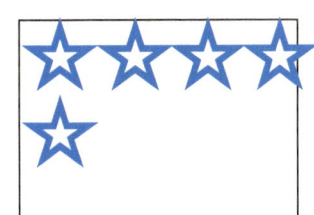

iv.

 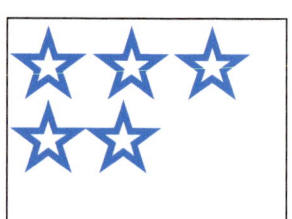

v.

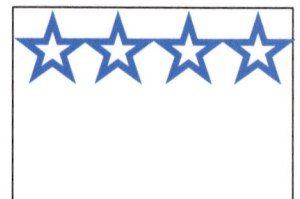

vi.

8. Match the stars with the correct numbers

5	☆☆
4	☆☆☆☆☆
3	☆
2	☆☆☆
1	☆☆

9. Complete the table below

1	2	3	4	5

10. Addition and subtraction Exercises

4	5	3
+2	+1	-2
□	□	□
8	6	9
-2	-3	-7
□	□	□
6	7	3
-1	+6	+3
□	□	□
8	7	6
-5	+2	-0
□	□	□
8	9	7
-1	-2	+4
□	□	□

About the Author

David Itanola has taught mathematics at various levels for more than 25 years. This book is the result of his extensive research on quick and easy strategies for teaching math to kids. Apart from writing on mathematical subjects, David has a penchant for helping people living a productive life.

David holds the master's degree in Mathematics. He is married with three children and enjoys playing the piano and mentoring children.

www.ingramcontent.com/pod-product-compliance
Lightning Source LLC
Chambersburg PA
CBHW040234220526
45473CB00001B/238